Using Math in Science

by Kathy Furgang

Table of Contents

2
20
6
How is scientific information obtained, organized, displayed, and analyzed?
18
10%
25%
20
32

INTRODUCTION

Math and Science in Our World Today

What would your life be like without electricity? What would the world be like without cars or jet planes? Picture a world without television or video games. What if you weren't able to keep track of time? It's easy to check a clock for the time and get on with your day. Without math and science, none of these things would exist.

In time, people gained scientific knowledge. They learned about the materials that make up life. They learned how to build things. They learned how to send messages through the air. They learned how to make a plane fly through the sky. Their lives became easier with each invention and discovery. One language made all of this possible.

That language is mathematics. We use math to measure. We use math to describe things. We use math to study information. We also use math to share information. Together, math and science have helped us cure deadly diseases. Math and science have helped people go to outer space. Math and science give us the tools to predict weather, stay safe, and save lives.

We even rely on math and science for fun! Think about going to the movies. Science and math help people make movies. We use science and math to project a movie onto the big screen. Science and math even help make your popcorn taste just right!

Who knows what new advances math and science will bring us in the future? Whatever they may be, the language of math is worth learning!

Each year, advances in math and science influence design and technology.

measurement

How are quantities measured and described?

Using measurements is one way a scientist can show data. All experiments need to be repeated. shows that the results can be cated. Numbers and quantities are important for this to happen. Just try plain the results of an experiment out the language of math. Your nunication would be difficult. uld also be inaccurate.

We use units to measure amounts. Quantitative data (KWAN-tih-tay-tiv DAY-tuh) tell about quantities. We use numbers to show quantitative data. For instance, we use numbers when we talk about length. We call data without numbers qualitative data (KWAH-lih-tay-tiv DAY-tuh). We use qualitative data to describe what something looks like. Scientists use both types of data in their work.

21.64 m

legant Louvre Pyramid ris, France, is made of and glass. It was built een 1985 and 1989. ase is 35.42 meters eet) wide. The height of yramid is 21.64 meters et) tall.

- customary system p. 8
- dimensional analysis p.
- estimate p. 16
- exponent p. 14
- formula p. 12
- metric system p. 9
- scientific notation p. 14

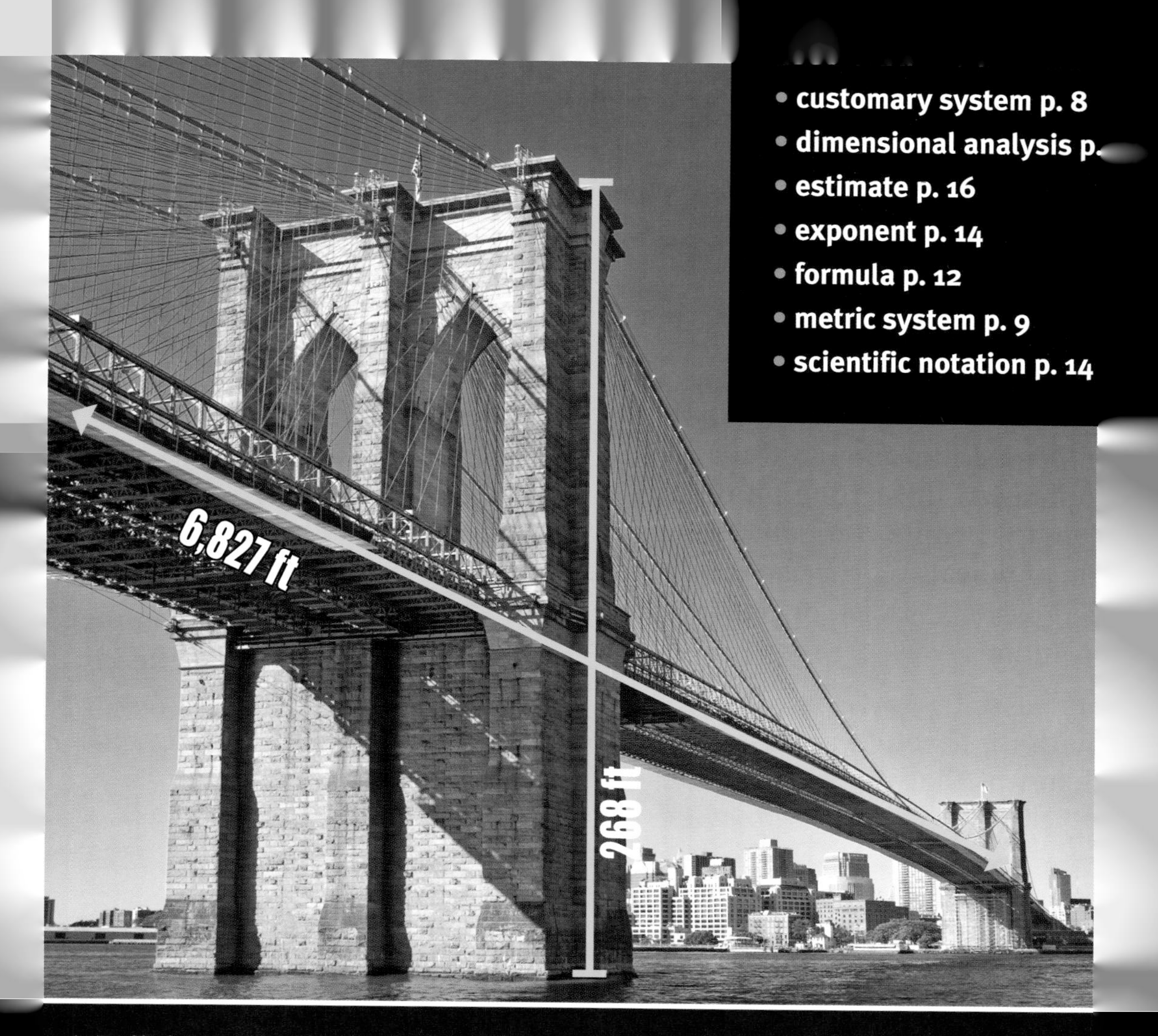

▲ Built between 1870 and 1883, the Brooklyn Bridge in New York is a remarkable achievement in engineering. As beautiful as it is sturdy, the bridge has a total length of 6,827 feet (2,081 meters). The height of each tower is 268 feet (81.7 meters) above high water.

e magnificent Temple of etzalcoatl in Chichen Itza, xico, was built between and 900 C.E. This massive ne structure is 24 meters .7 feet) in height, not luding the 6-meter .7-foot) temple on top. ▶

The Customary System

Long ago, people began to use measurements. This helped people talk about things like height and length. At first, people used body parts to measure. An inch was the width of a thumb. A foot was the length of a foot, and so on. A mile was about a thousand paces. People used these rough units to describe length. People compared things to amounts of wheat to describe weight. People used baskets, sacks, or pots to describe volume.

As years passed, people saw the problem with these rough units. The problem was that these units were not exact. People's thumbs and feet are different sizes. This meant the length of an inch or foot was never the same. People saw that they needed standard units. First they standardized the exact length of an inch. Then they defined a foot as exactly 12 inches. Next they defined a yard as exactly 3 feet, and so on. Today we call these standardized measurements the **customary system** (KUS-tuh-mair-ee SIS-tem).

Still, a few problems remained. Different countries used different standard units. Also, changing feet to inches, or pounds to ounces, was not simple.

BODY PARTS AND ANCIENT MEASUREMENT

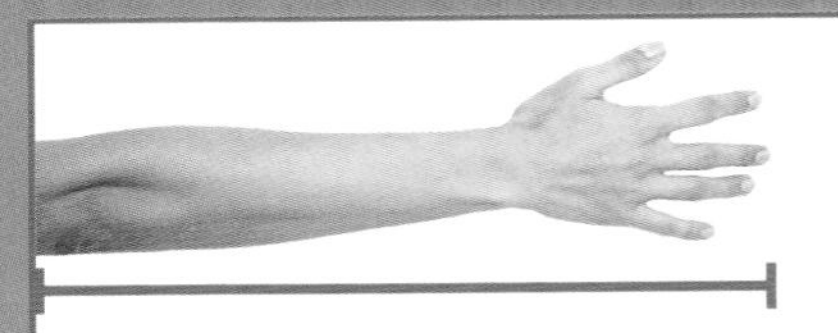

Cubit
Ancient = from the elbow to the end of the fingertips
Today = 18 inches

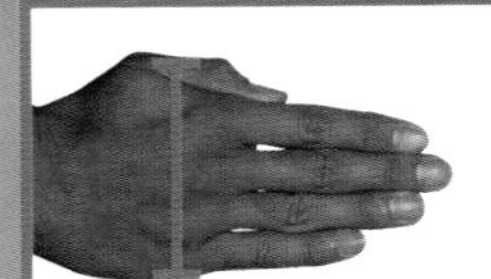

Hand
Ancient = 5 digits
Today = 4 inches (horses are measured in hands)

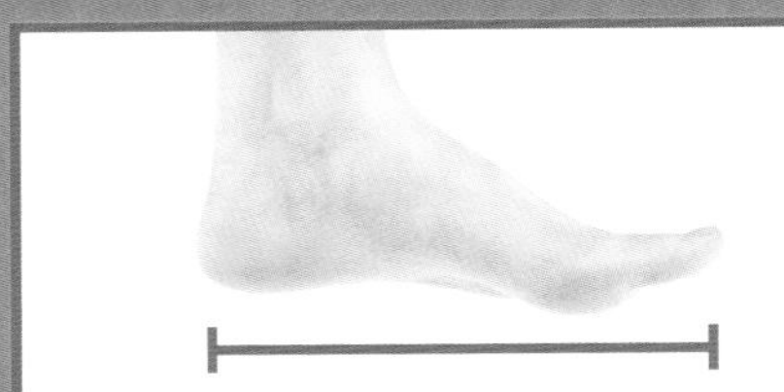

Foot
Ancient = 11 1/42 inches
Today = 12 inches

Yard
Ancient = distance from a man's nose to the end of his outstretched arm
Today = 36 inches

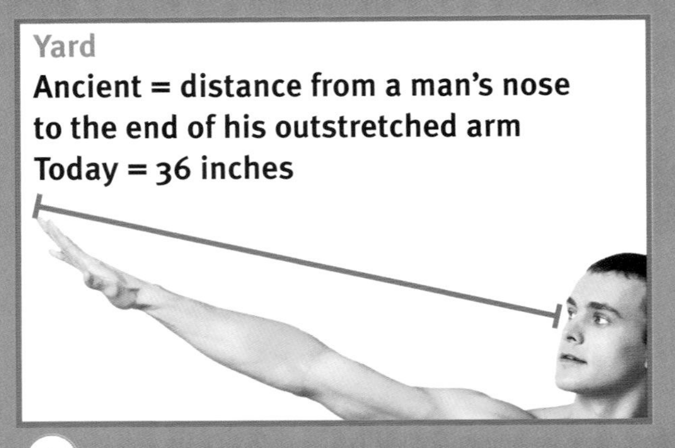

Mile
Ancient = 1,000 paces
Today = 5,280 feet

Quantity	SI Base Unit	Metric	Customary
length	meter (m)	millimeter (mm)	inch (in)
		centimeter (cm)	foot (ft)
		meter (m)	yard (yd)
		kilometer (km)	mile (mi)
mass	kilogram (kg)	gram (g)	n/a
		kilogram (kg)	n/a
weight	n/a		ounce (oz)
		newton (N)	pound (lb)
volume	n/a	milliliter (mL)	fluid ounce (fl oz)
		liter (L)	gallon (gal)
temperature	kelvin (K)	Celsius (°C)	Fahrenheit (°F)
time	second (s)	second (s)	second (s)

The Metric System

In the late 1700s, people thought of a new system to avoid these problems. We call this system the **metric system** (MEH-trik SIS-tem). France adopted these units first in the 1800s. Soon after, other nations also accepted the metric system. The metric system is a standardized system of measurements. The units are based on multiples of 10. Using the number ten makes conversions easy. Even better, the measurements are based on fixed standards. The metric unit of length was based on the distance around Earth. Then the units for volume and mass were derived from this length. This way, the basic units were related to one another. Larger and smaller multiples of these units could be obtained simply by moving the decimal point to either the right or to the left.

You can also tell if a measurement is large or small by looking at the prefix. A meter is a base unit of length. The prefix *centi-* means "one-hundredth." A centimeter is one-hundredth of a meter. On the other hand, the prefix *kilo-* means "one thousand." So, a kilometer is one thousand meters.

Most countries use the metric system. Some countries still use the customary system. So, scientists around the world needed one universal system. This would allow them to share exact measurements and data. In English, this system is called the International System of Units (SI). SI has seven base units. The meter is the unit for length. The kilogram is the unit for mass. The second is the unit for time. The kelvin is the unit for temperature. The SI also has three more base units that chemists and physicists use to share research.

HANDS-ON SCIENCE
Measuring Volume with Accuracy

MATERIALS

- 5 graduated cylinders ranging in size (10 mL, 25 mL, 50 mL, 100 mL, 250 mL)
- 25-mL beaker

TIME

45 minutes

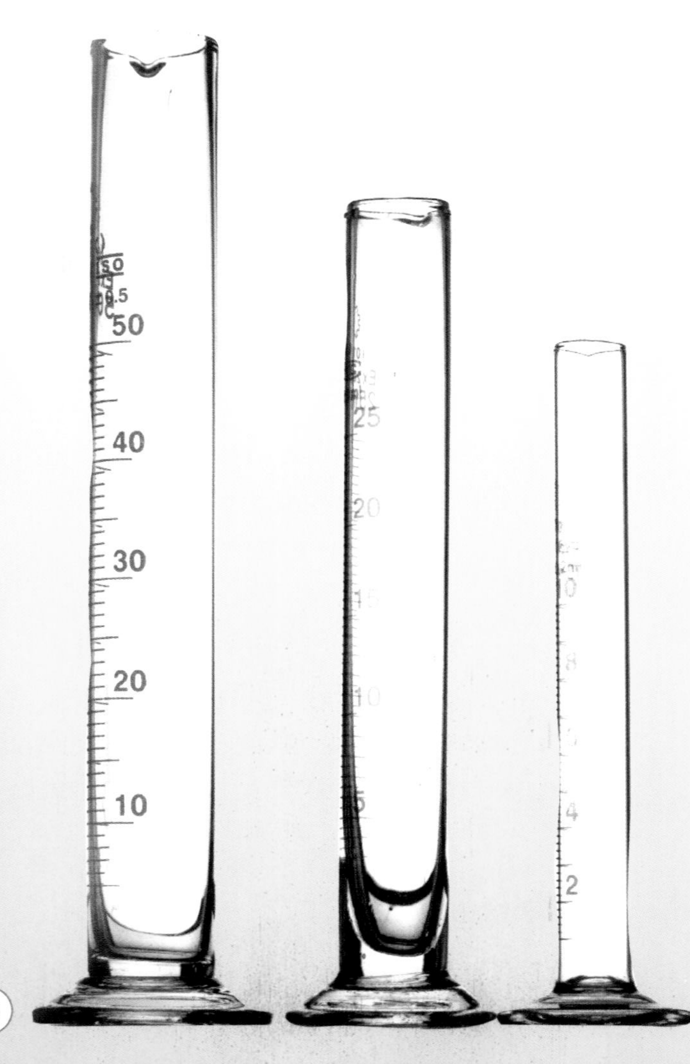

PROCEDURE

1. Fill the beaker with water.
2. Transfer 5 mL of water from the beaker to a 250-mL graduated cylinder.
3. Then pour the 5 mL of water in the 250-mL graduated cylinder into a 10-mL graduated cylinder.
4. Carefully read and record the amount of water seen in the 10-mL graduated cylinder.
5. Repeat steps 1–3 three more times with the remaining graduated cylinders.
6. Compare the four final measurements recorded in step 4.

ANALYSIS

1. Describe how you would measure 15 mL of water needed for an experiment.
2. What general statement (rule) might you make about the laboratory equipment used in an experiment?

Length

Length is how long something is. The SI base unit for length is the meter. We can use a meter stick to measure length. We can also use a metric ruler.

Temperature

Temperature is the amount of energy in the particles of matter. The base SI unit for temperature is degrees kelvin. Most scientists also accept degrees Celsius (SEL-see-us). On the Celsius thermometer, water freezes at 0°C. Water boils at 100°C. We can use a thermometer to measure temperature.

Volume

Volume (VAHL-yoom) is the amount of space something takes up. SI has no base unit for volume. Most scientists measure volume in liters or milliliters. We can use graduated cylinders to measure the volume of liquids.

Mass

Mass is the amount of matter that an object has. The SI base unit for mass is the kilogram. We can use a balance to measure mass.

▼ We can use a triple beam balance to measure mass.

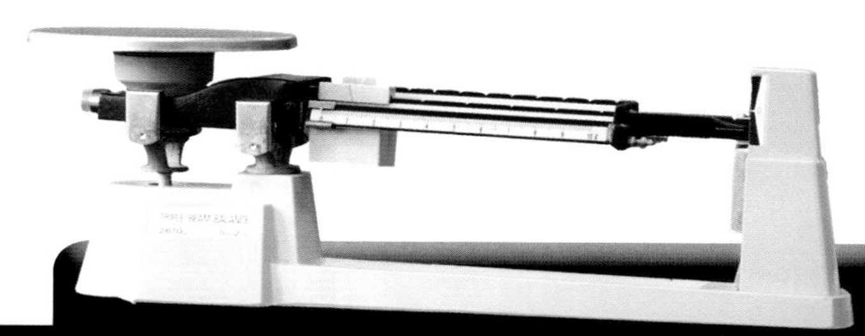

EVERYDAY SCIENCE

Mass vs. Weight

People often confuse the measurements of mass and weight. Mass is the amount of matter in an object. Weight is the measure of gravity's pull on that object. A balance tells the mass of an object, and a scale tells the weight of an object. To tell the difference, remember that your mass is always the same wherever you go. You have the same amount of matter in you. But if you go to the moon, your weight will be different because the pull of gravity on the moon is different from the pull of gravity on Earth.

Derived Units

We call some units derived units. We use more than one measurement to find a derived unit. Speed is a derived unit. First you need to measure how far you've traveled. Then you need to measure how long the trip took. So, you need to measure distance and time. Then you need to use a formula. A **formula** (FOR-myuh-luh) is a mathematical rule or relationship that is shown in symbols. We can use this formula to find speed:

$$S = \frac{d}{t}$$

$$\text{Speed} = \frac{\text{distance}}{\text{time}}$$

Volume can be a derived unit. Think about how you would find the volume of a rectangular solid. First, you measure its length, width, and height. Then you multiply the measurements ($v = l \times w \times h$). Density is also a derived unit. You can use a formula to find density. The formula used to find density is:

▼ The bike rider's speed can be found by dividing the distance traveled by the time it took to travel that distance.

SCIENCE AND TECHNOLOGY

Accuracy vs. Precision

What is the difference between measurements that are accurate and ones that are precise? These may sound very similar based on the dictionary definitions, but they are different in science. In science, measurements must be both accurate and precise in order for results and measurements to be repeated. When a measurement is accurate, it is correct. But does it mean that it is precise? A precise measurement is one that can be repeatedly determined with the same accuracy each time. You must repeat measurements many times to see if they are precise. You may find that the average of your measurements is accurate, but each individual measurement is off by a small amount. Repeating measurements to be sure they are both precise and accurate is an important part of science.

Checkpoint ✔

Talk It Over

The United States is one of the only countries in the world that uses the customary system of measurement. Europe, Asia, and the rest of the world use the metric system exclusively. Do you think the United States should convert to the metric system? Why or why not?

Unit Conversions

You can convert, or change, units from one to another. When you convert units you use **dimensional analysis** (dih-MEN-shuh-nul uh-NA-lih-sis). This is a math concept. The idea is that you can show the same measurement in different ways. When you convert yards to inches, you are not changing the amount of length. You are simply changing the unit used to measure the length.

You can use dimensional analysis in the metric system. You can convert 500 milliliters to 0.5 liter. You can also convert time into different units. How many seconds are in 24 hours? You can multiply to find out. First, multiply 24 hours by 60 minutes. Then multiply the total by 60 seconds. The final total is 86,400. Therefore, 24 hours is equal to 86,400 seconds.

You can use dimensional analysis in the customary system. You can convert yards to feet. 1 yard is 3 feet. Then you can divide to find how many yards are in 108 feet.

You can even convert customary units to metric units or SI units. Look at the conversion table on page 34. You can use this table to know what number to multiply or divide by.

Sometimes you do not need to multiply or divide. Scientific rulers have metric units and customary units. Look at the ruler below. You can see that there are 2.54 centimeters in 1 inch.

▲ Some rulers allow you to convert a customary unit to a metric unit without calculations.

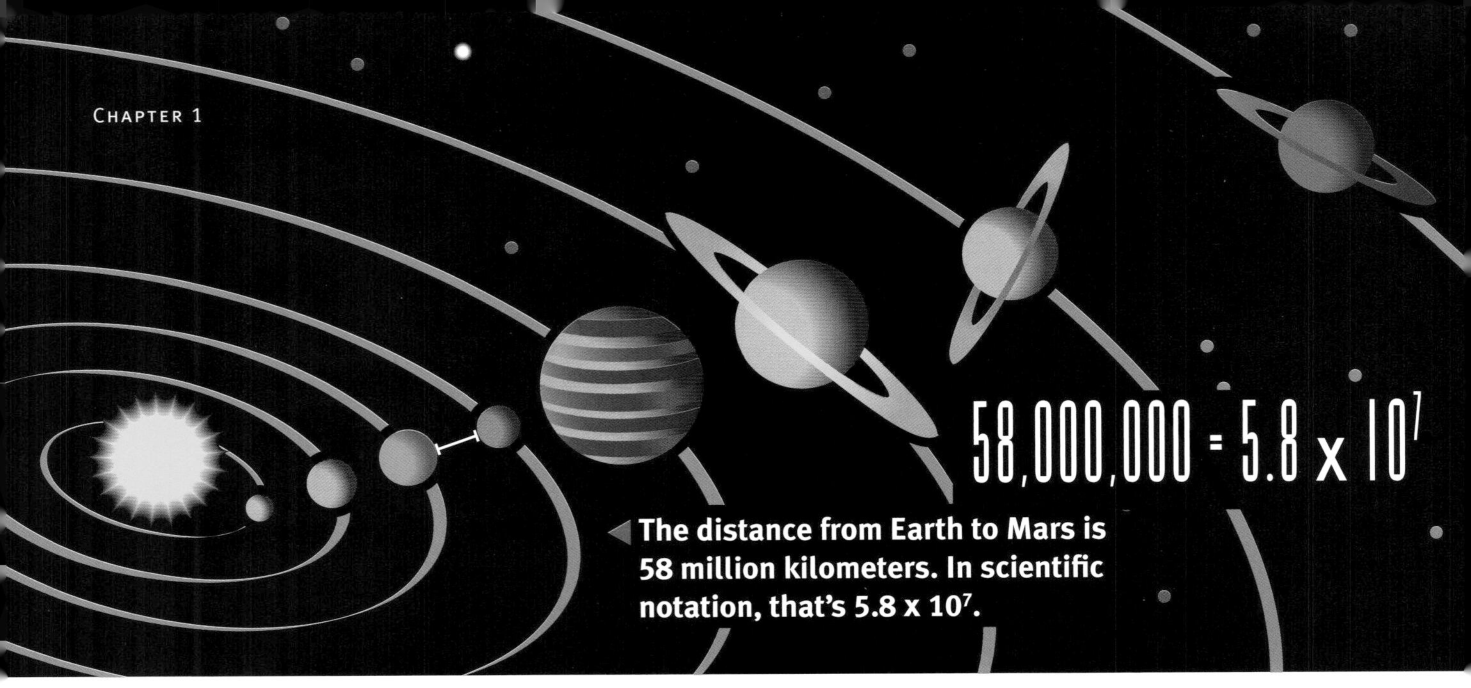

The distance from Earth to Mars is 58 million kilometers. In scientific notation, that's 5.8×10^7.

Writing Numbers in Scientific Notation

Some numbers are too big or too small to write out. For example, 4 trillion looks like this: 4,000,000,000,000. We can use **scientific notation** (sy-en-TIH-fik noh-TAY-shun) to write this number more easily. Scientific notation is a way to write numbers using exponents. An **exponent** (EK-spoh-nent) is a number we write to the upper right of a number. An exponent tells us how many times to multiply a number by itself. For example, $10^3 = 10 \times 10 \times 10$, so $10^3 = 1{,}000$ or one thousand. One trillion is equal to 10^{12}. So, 4 trillion in scientific notation is 4×10^{12}.

Scientific notation works with decimals, too. For example, $10^{-3} = 10 / 10 / 10$, so $10^{-3} = 0.001$ or one-thousandth. We can use this notation to show the age of a fossil or the mass of an atom.

Look at the chart on page 15. This chart shows the pattern of scientific notation.

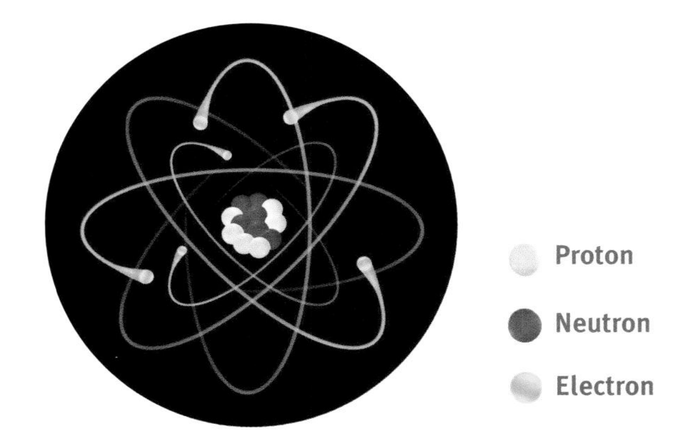

$$m = 9.10938188 \times 10^{-31} \text{ kg}$$

▲ **The mass of an electron is 0.000000000000000000000000000000910938 kg. That is 9.10938×10^{-31} kg in scientific notation.**

SCIENCE AND MATH

Scientific Notation

A unit of measurement often used when describing atoms is a nanometer. One nanometer (nm) is one billionth of a meter. How many meters is 15 nanometers? Use scientific notation for your answer.

Number in Words	Decimal	Scientific Notation
one-thousandth	0.001	1×10^{-3}
one-hundredth	0.01	1×10^{-2}
one-tenth	0.1	1×10^{-1}
one	1	1×10^{0}
ten	10	1×10^{1}
one hundred	100	1×10^{2}
one thousand	1,000	1×10^{3}

▼ Scientists can date rocks and fossils back to the time they were formed. The oldest rock on the Earth's surface is more than 3,800,000,000 years old. That's 3.8 billion, or 3.8×10^9 years old.

$$3{,}800{,}000{,}000 = 3.8 \times 10^9$$

Estimation

Sometimes you do not need an exact amount. Sometimes an **estimate** (ES-tih-mit) is good enough. An estimate is not exact, but it is close to the actual amount. The speed of light is 299,792.458 kilometers/second (186,411.358 miles/second). You might round that number up. You can round that number to 300,000 kilometers (185,000 miles) per second. You can use this number to make an estimate. But if you are doing an exact calculation, your answer would be wrong.

When doing an experiment, you might estimate the results. But the measurements you record and the calculations you make must be based on exact measurements.

Can you estimate the number ▶ of coins in this jar?

Summing Up

- Long ago, people established units and systems to measure length, mass, volume, temperature, and time in a consistent way.
- The customary system (inches, pounds, gallons, etc.) is the standardized system of units used in the United States.
- The metric system is based on multiples of ten. These standardized units (meters, liters, grams) are used in most countries around the world.
- We can express measurements in different units in the same system or convert back and forth between systems of measurement.
- Scientists use a universal system of measurement to share their research called the International System of Units (SI). The meter (length), the kilogram (mass), the second (time), and the kelvin (temperature) are among the seven universal or SI units.
- Derived quantities, such as speed and density, are calculated using formulas.
- Measurements can be expressed in shortened forms. They can also be expressed as estimates or exact numbers, depending on how they will be used.

Putting It All Together

Choose from the research activities below. Work independently, in pairs, or in small groups. Then share your findings with the class.

1. Throughout history, people have used a variety of units of measurement. In a group, research early or unusual units of measurement, such as units of barley or salt. Present your findings in a poster.

2. Work with a partner to make a diagram of your classroom. Show measurements, such as the length of the walls, the board, and the desks. Include units for each measurement.

3. Research distances between planets and the sun in kilometers. Create a poster that shows the arrangement of planets and describes their distance from the sun in scientific notation.

If the Shoe Fits...
Cartoonist's Notebook • Illustrated by Joel Carlson
MATH CLASS
Before I go into the details of my report on "Shoe Sizes Across the World," let's talk about standard measurement.
Yes, let's!
Before measurement became standardized, an inch was roughly the width of a thumb.
But as you can see from my two assistants here, there is a difference from thumb to thumb.
OW!
Hey, watch the nails.
A foot was the length of the king's foot, which is why we call this measurement a ruler.
So I have small feet! My mother says they are dainty!
But as you can see, again, from my two assistants here, there is a difference from foot to foot.
Big feet run in my family.
I can see.
And smell
the difference.

The International System of Units (SI) was established in the 1960s. Why do you think it's important to have a universal system of standard measurements? How does this help scientists?

What are some other systems that you would like to make universal?

CHAPTER 2

Mathematical Relationships

How can relationships between numbers be described and determined?

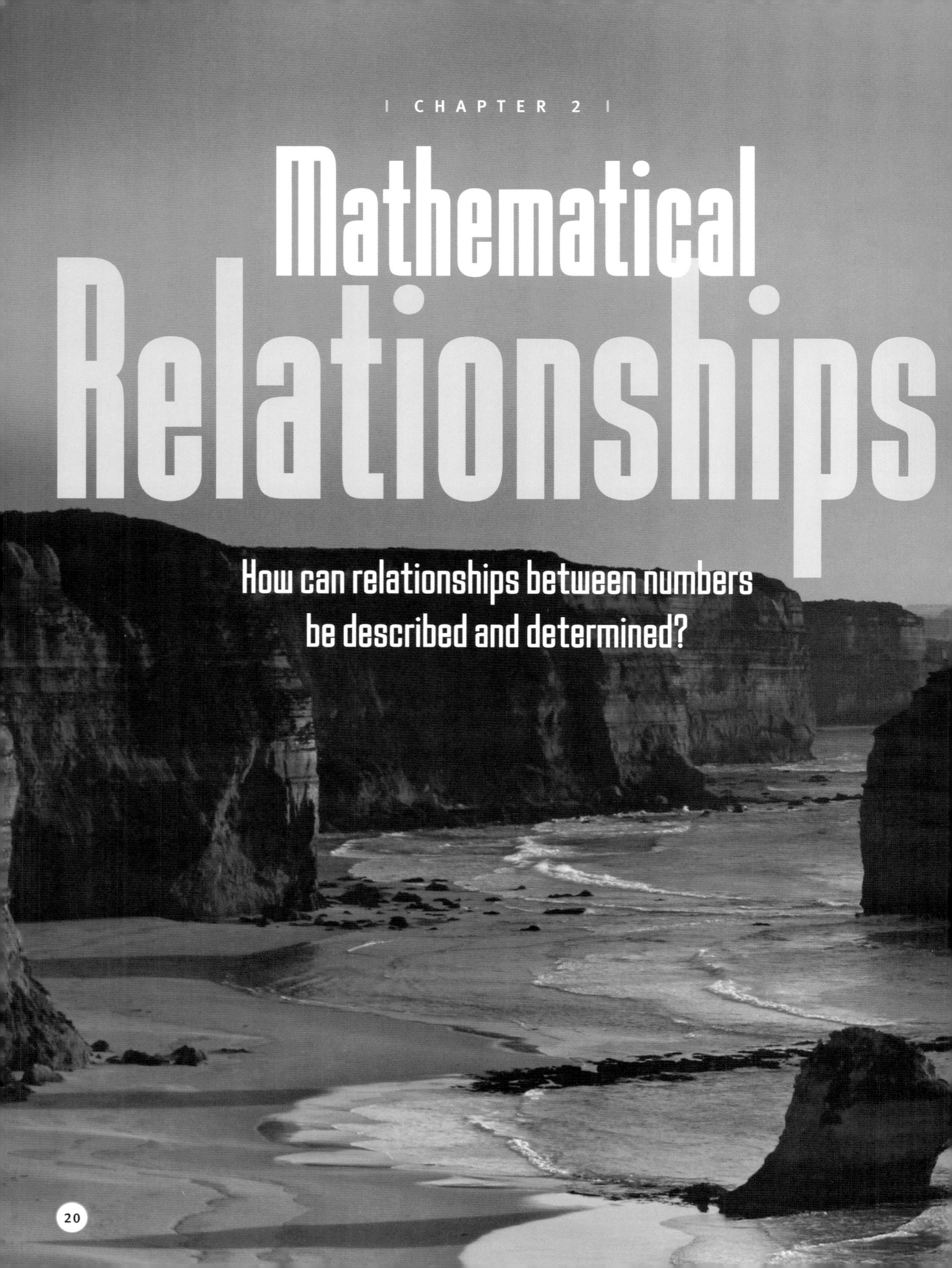

We have ten digits in all: 0, 1, 2, 3, 4, 5, 6, 7, 8, and 9. We use these digits to write numbers. We can calculate with numbers. We can communicate with them. We can also write the same numbers in different ways. Scientists can use numbers to show data in different ways.

ESSENTIAL VOCABULARY

- **decimal p. 23**
- **denominator p. 22**
- **fraction p. 22**
- **improper fraction p. 22**
- **mean p. 30**
- **median p. 30**
- **mode p. 30**
- **numerator p. 22**
- **percent p. 23**
- **proportion p. 25**
- **range p. 30**
- **rate p. 24**
- **ratio p. 24**

◀ **Recent studies show that 66% of California's coastline is actively eroding. You can use a fraction, decimal, or percentage to express this number. You can also use rates to describe the amount of erosion per year.**

$$66\% = 0.66 = \frac{2}{3}$$

1/2 1/3 1/4 1/5 1/6 1/8

1

Writing Numbers in Different Forms

We have three main types of numbers. Whole numbers are numbers greater than or equal to one. We can use fractions, decimals, or percents to show numbers less than one.

Fractions

A **fraction** (FRAK-shun) is a number that shows a part of a whole. All fractions have a **numerator** (NOO-muh-ray-ter). The numerator is the amount above the line. Fractions also have a **denominator** (dih-NAH-mih-nay-ter). The denominator is the amount below the line. The denominator tells the total number of equal parts. The numerator tells how many equal parts are being counted.

Five-eighths (5/8) is a fraction. This fraction shows 5 of 8 equal parts. Five-eighths is less than one. Eight-eighths (8/8) is equal to one. Nine-eighths (9/8) is greater than one. We call this an **improper fraction** (im-PRAH-per FRAK-shun). An improper fraction is greater than one. Nine-eighths (9/8) can also be written as the mixed number one and one-eighth (1 1/8).

Decimals

A **decimal** (DEH-sih-mul) is also a number that shows parts of a whole. The decimal system is based on units of ten. We can use decimals to show fractions. 1/10 is equal to 0.1. That means that 0.1 shows 1 of 10 equal parts. We can use decimals to show money. We can also use decimals to show measurements.

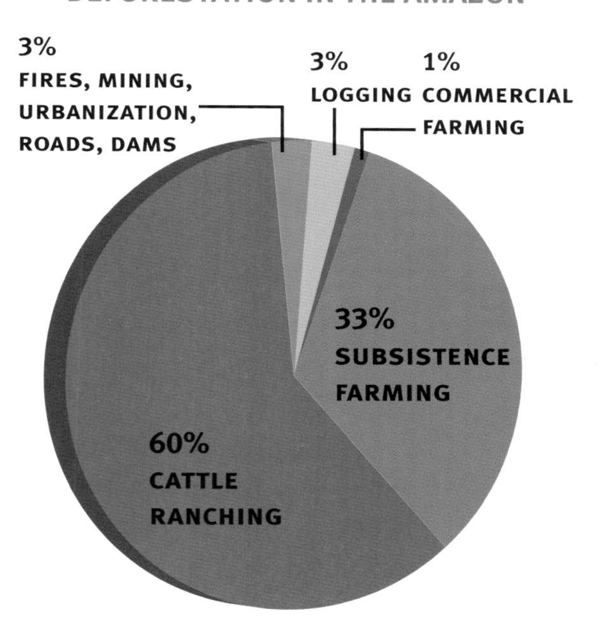

▲ Data results are often expressed in percents.

Percents

A **percent** (per-SENT) is a number that shows the number of parts out of 100. We can use percents to show different types of data. A scientist can convert data to percents. A scientist can also use percents to study and share results.

▼ We can show equivalent amounts in the form of fractions, decimals, and percents.

$$\frac{1}{10} = \frac{10}{100} = 0.1 = 10\%$$

$$\frac{1}{4} = \frac{25}{100} = 0.25 = 25\%$$

$$\frac{1}{2} = \frac{50}{100} = 0.5 = 50\%$$

Relationships Between Numbers

We can relate numbers to one another. Numbers in a data set can have different relationships. These relationships can mean different things.

Ratios

A **ratio** (RAY-shee-oh) is the relationship between two amounts. We can write the ratio of x to y as $x{:}y$, or x/y. If you have 4 short-stemmed plants and 8 long-stemmed plants, the ratio would be 4:8 or 4/8. You can simplify this to 1:2. The ratio says for every 1 short-stemmed plant, you have 2 long-stemmed plants. We can use ratios when looking at surveys.

▼ A ratio is a good way to express relationships between quantities.

4:8 = 1:2

SCIENCE AND MATH

Ratios in Chemistry

In a compound you can find the ratio of combining atoms by looking at the subscripts in the chemical, or molecular, formula. The formula for water is H_2O. That means the ratio of hydrogen atoms to oxygen atoms is 2 to 1, 2:1, or 2/1.

Rates

A **rate** (RATE) is a special type of ratio. A rate shows the relationship between amounts with different units. We use rates to describe speed. 90 kilometers (56 miles) per hour is a rate. The word *per* is often a signal word that a rate is being used. We can also show rates as x/y.

▼ A rate is often used to show speed.

Proportions

A **proportion** (pruh-POR-shun) is the relationship between two equal ratios. We can use proportions to make models or maps. Let's say every centimeter on a map is equal to 5 actual kilometers. You have a 1:5 ratio. So, every 5 centimeters on the map will be equal to 25 actual kilometers. The proportion can be shown as:

$$\frac{1}{5} = \frac{5}{25}$$

The Root of the Meaning: The word

PROPORTION

is based on the Latin word *proportio*, which means relationship of parts or analogy. An analogy is a comparison between two related things.

When you make a scaled model of an object, you use proportions. Suppose you were making a model of a human cell. You want to know what size to make each cell part. The diameter of a human cell is 0.01 cm. The diameter of your model cell is 100 cm. The ratio is 0.01:100. The nucleus of a human cell is 0.0005 cm in diameter. You can find the diameter of the nucleus for your model by solving the proportion.

$$\frac{0.01}{100} = \frac{0.0005}{?}$$

▲ You can make an accurate model to scale by solving the proportion between the parts of the actual object and the parts of the model.

Direct and Inverse Relationships

Graphs can help you see the relationship between variables. A direct relationship is when a change in one variable causes the same change in another. If one variable goes up, the other goes up. If one variable goes down, the other goes down. We call this a direct correlation (kor-uh-LAY-shun).

Here is an example. If the number of times a wheel rotates goes up, the distance the wheel travels will also go up. The number of plants has a direct relationship with animals in an ecosystem. So an ecosystem with many plants will have many animals. An ecosystem with few plants will have few animals.

On a line graph, a direct relationship has a positive slope. That means the line would move up in a straight line as traveled along the *x*-axis and *y*-axis.

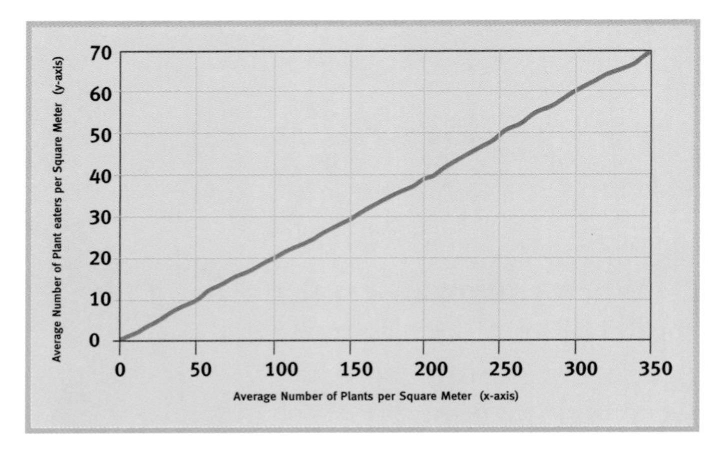

▲ **There is a direct relationship between the number of plants and plant eaters in an ecosystem.**

An indirect, or inverse relationship is the opposite. Here a change in one variable causes the opposite change for another. If one variable goes up, the other goes down.

Here is an example. The greater an oil spill, the lower the amount of marine life in that area. We call this type an indirect correlation.

On a line graph, an indirect relationship has a negative slope. That means the line would move down in a straight line as traveled along the x-axis and y-axis.

Checkpoint

Visualize It

How can you use art to describe a direct and an inverse relationship? You might use a seesaw to see the concept more clearly. In an inverse relationship, when one side of the seesaw goes up, the other goes down. This is like the slope you might see on a line graph.

There is an inverse relationship between the size of an oil spill and the population of seabirds in a region.

Formulas

We use formulas all the time in science. Formulas show relationships between amounts. Formulas help us find volume. Formulas also help us find area. We can also use them to find speed, density, or force. We even use formulas to convert temperature to Kelvin.

We use letters to show variables in a formula. Albert Einstein was a famous scientist. Albert Einstein's formula is $E = mc^2$. In this formula, E represents energy, m means mass, and c is the speed of light in a vacuum. When the letters are replaced with the appropriate values for two of the three variables, the third variable can be found.

Albert Einstein came ▶ up with one of the most famous mathematical formulas of all time, $E = mc^2$.

To use any formula, you need to know certain measurements. Look at the formula for density below. First plug the numbers you know into the equation and then solve for the variable that you do not know.

▼ The table below lists the formulas for relationships frequently explored in math and science.

Measurement	Relationship	Formula
area of a rectangle	area = length x width	$A = l \times w$
area of a triangle	area = 1/2 (base x height)	$A = (b \times h)/2$
density	density = mass/volume	$d = m/v$
volume of a rectangular prism	volume = length x width x height	$v = l \times w \times h$
speed	speed = distance/time	$S = d/t$
acceleration	acceleration = (final velocity – original velocity)/time	$A = (fv - ov)/t$
force	force = mass x acceleration	$F = m \times a$

SCIENCE AND MATH

Find the speed of an object that travels a total distance of 5 km in 15 minutes. Use the appropriate formula from the chart above.

$$\text{Density} = \frac{\text{mass}}{\text{volume}}$$

Measures of Central Tendency

Sometimes you have a series of numbers in a data set. Finding the mean, median, mode, and range can help you understand the data.

Mean

The **mean** (MEEN) of a set of numbers is the average. The mean is the sum of the numbers divided by the number of items. Let's say you have the following set of numbers: 17, 16, 21, 18, 18, 15, 12. The mean would be 16.7.

17 + 16 + 21 + 18 + 18 + 15 + 12 = 117/7 = 16.7

Median

The **median** (MEE-dee-un) is the middle of the set. About half the numbers are above and half the numbers are below the median. (For an even set of numbers, average the two middle numbers to find the median.) The median for this set is 17.

12 15 16 17 18 18 21

Mode

The number that appears most often in a set is the **mode** (MODE). It helps to know the number that occurs most frequently in a set. Sometimes the mode and the mean can be the same number. 18 appears twice, so the mode is 18.

12 15 16 17 18 18 21

▲ **Measures of central tendency can be used on a wide variety of scientific data.**

Range

The **range** (RANJE) is the difference between the lowest and highest numbers in the set. The range tells you how spread out the data is. The range of this set is 9.

12 15 16 17 18 18 21   21 − 12 = 9

SCIENCE AND MATH

Look at the Celsius temperatures on the climograph below. Find the mean, median, mode, and range of the data: −8°, −5°, −4°, 0°, 10°, 12°, 13°, 13°, 9°, 5°, −1°, −7°

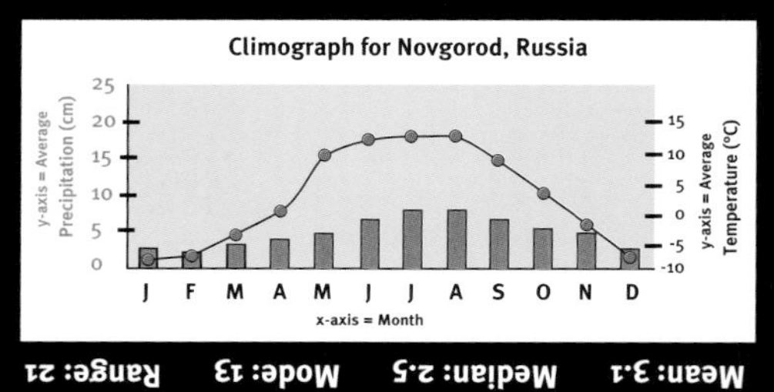

Mean: 3.1   Median: 2.5   Mode: 13   Range: 21

Summing Up

- We can use numbers to show different relationships between numbers.
- When we represent numbers as fractions, decimals, or percents, they can be converted back and forth easily to represent a part of a whole.
- Ratios, rates, and proportions also show how numbers can be related to one another.
- Direct and inverse relationships as well as formulas can show that numbers represent real data.

Putting It All Together

Choose from the research activities below. Work independently, in pairs, or in small groups. Then share your findings with the class.

1 Do a survey of the footwear worn by your class. Find what percent of students is wearing each type of shoe (sneakers, sandals, boots, or others). What do you notice about the sum of the percents?

2 Write a sentence explaining what a rate is. Then give a rate and tell what it describes. Indicate what an increase or decrease in the rate indicates. Compare your example with those of your classmates.

3 Find the amount of rainfall in your city or town each month last year. Round each amount to the nearest whole number. Then find the mean, median, mode, and range of the data. Suppose the greatest amount is doubled. How would this change affect the measures of central tendency?

CHAPTER 3

Representing and Interpreting Data

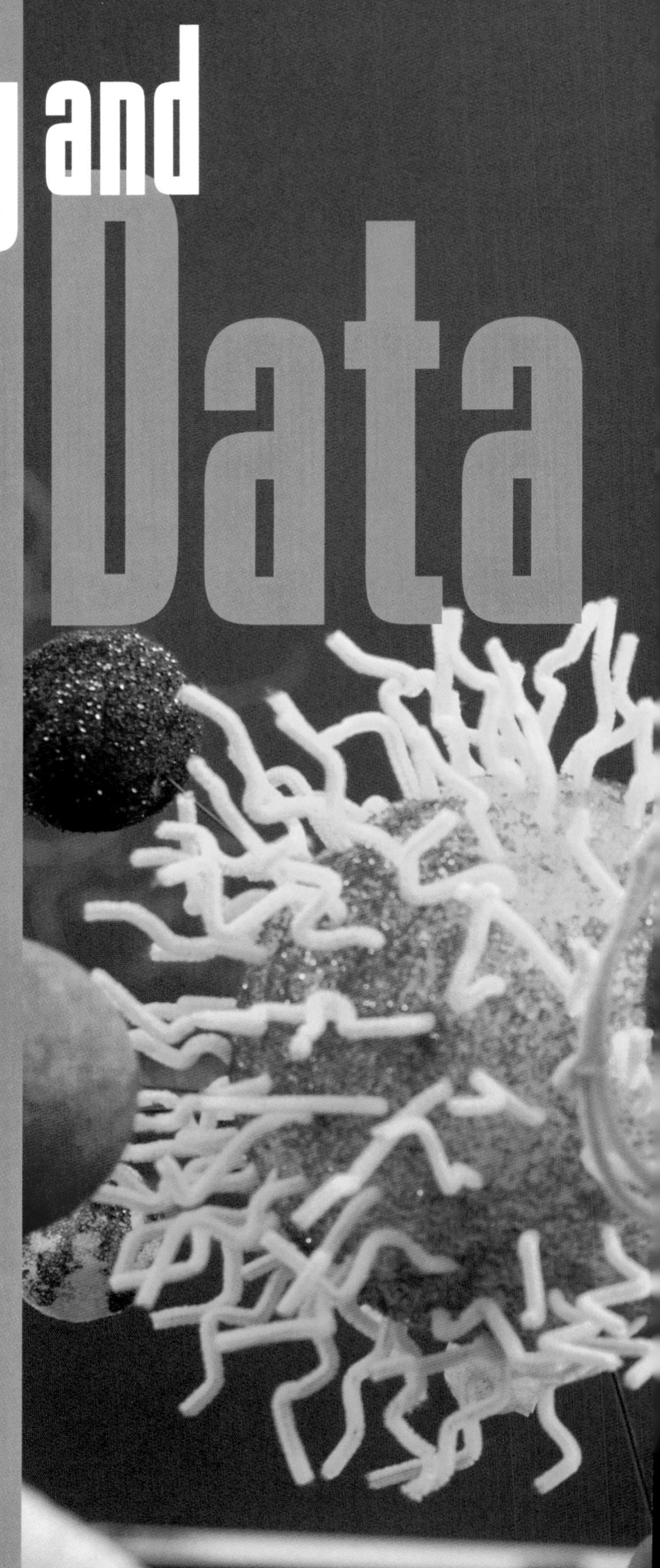

How can data be organized and interpreted?

During an experiment, you must record data properly. When sharing your results, you must also show data clearly. Other scientists can then compare your findings with their own work.

You can use tables and graphs to show results quickly and clearly. Charts and data tables can help us organize our information. This makes it easier to study and understand.

ESSENTIAL VOCABULARY

- bar graph p. 35
- circle graph p. 39
- histogram p. 38
- line graph p. 36
- scatter plot p. 40

Data Tables

A data table is a simple way to show data. A data table has rows and columns. Each column or row has a title. The table itself also has a title.

Suppose you want to investigate how far a toy car will roll off a ramp. You decide to repeat the trial seven times. This allows you to get a data set to analyze. A data table is a useful tool for recording this information. Look at the data table to the right. The table has a title. Each column also has a title. You can see the unit of measure (meters). The display makes it easy to compare the numbers in the set.

Distance Car Traveled From Ramp

Trial Number	Distance Traveled (meters)
Trial 1	14
Trial 2	17
Trial 3	12
Trial 4	15
Trial 5	14
Trial 6	16.5
Trial 7	12.5

▲ Data tables help us record results.

▼ Conversion tables help us to convert metric units to customary units, and vice versa.

Measurement Conversion Table

Quantity	Metric		Customary	
length		1 millimeter (mm)	0.039 inch (in)	
	10 millimeters (mm)	1 centimeter (cm)	0.39 inch (in)	0.033 foot (ft)
	100 centimeters (cm)	1 meter (m)	3.9 feet (ft)	1.094 yards (yd)
	1,000 meters (km)	1 kilometer (km)	1,093.6 yards (mi)	0.621 mile (mi)
mass	1,000 milligrams (mg)	1 gram (g)	n/a	
	1,000 grams (g)	1 kilogram (kg)	n/a	
weight	0.1019 kilograms (kg)	1 newton (N)	3.597 ounce (oz)	0.2248 pound (lb)
volume	1 cubic centimeter (cm^3)	1 milliliter (mL)	0.0338 fluid ounce (fl oz)	0.2 teaspoon (tsp)
	1,000 milliliters (mL)	1 liter (L)	1.0566 quarts (qt)	0.264 gallon (gal)
temperature		0° Celsius (°C)	32° Fahrenheit (°F)	
time		1 second (s)	1 second (s)	

Bar Graphs

You can also show your data in a bar graph. A **bar graph** (BAR GRAF) shows different values with bars of different heights. The *x*-axis is the horizontal axis. The *x*-axis shows the trial number. The *y*-axis is the vertical axis. The *y*-axis shows the distance. Compare the bar graph with the data table. Which is easier to read? Why do you think so?

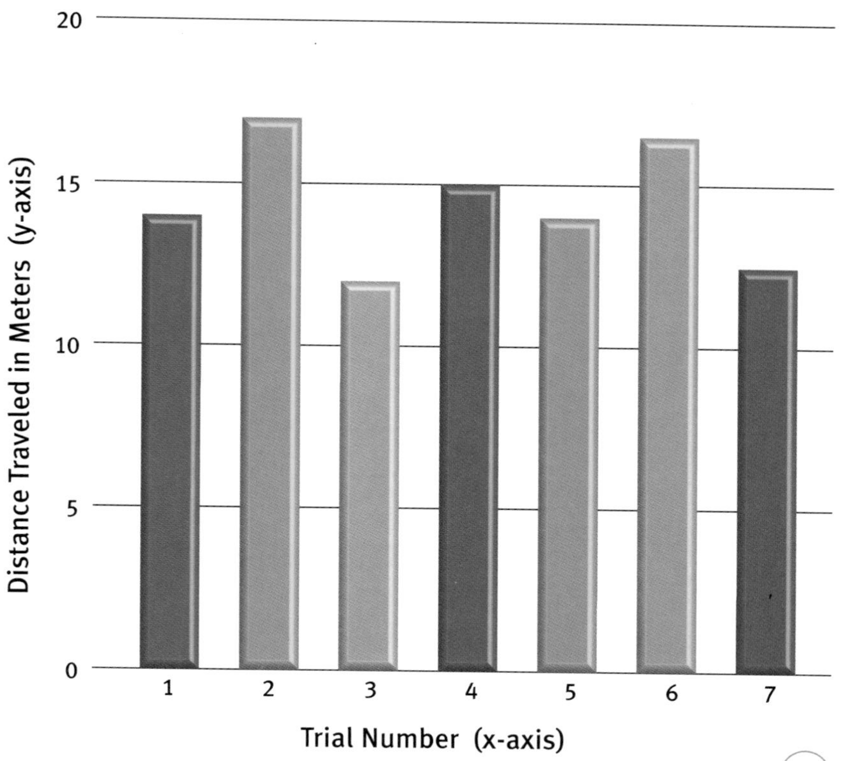

Line Graphs

A **line graph** (LINE GRAF) is a good way to represent data that changes over time. Like all graphs, a line graph should have a title.

The *x*-axis runs horizontally and shows one variable. The *y*-axis runs vertically and shows the other variable.

Look at the line graph below. This graph shows the number of eggs laid at a pond. The *x*-axis shows the month. The *y*-axis shows the number of eggs.

A line graph can show more than one set of data. You could show the number of eggs laid by fish as well. You just need to use a different color line. You must also use labels to explain what each line shows.

▼ The line graph shows data about when the ducks at this pond lay their eggs.

HANDS-ON SCIENCE
Make Your Own Graph

MATERIALS: graph paper, balloon, tape measure, felt-tip marker

1. Blow up the balloon and measure the circumference of the circle with a tape measure. Use a felt-tip marker to mark the exact part of the balloon that you measured. Record the measurement on a data table with the date.

2. The next day, measure the circumference of the balloon again. Record the measurement on the data table with the date.

3. Continue to measure the balloon circumference every day for five days.

4. Record your results on a line graph. Make the *x*-axis the number of days, and the *y*-axis the measurement of the balloon's circumference. What did your graph reveal about the balloon over time?

SCIENCE AND TECHNOLOGY

Graphing Calculators

A graphing calculator is a useful tool for scientists and mathematicians. It does more than just calculate equations. A graphing calculator can plot graphs and solve equations. It can also graph the next point on a line. If you enter the parameters into the calculator, you can enter a value for the *y*-axis and one for the *x*-axis, and a formula will be solved with that plot point placed on the graph.

Histogram

A **histogram** (HIS-tuh-gram) is a graph that shows the ranges in a data set. A histogram looks like a bar graph. The difference is the *x*-axis shows ranges. The *y*-axis shows the frequency of each range.

Here is an example. You may study a large sampling of people. These people may have different ages. A histogram can show you how many of your test subjects are in each age group. This type of graph works best when you have a large set of data and need to organize it.

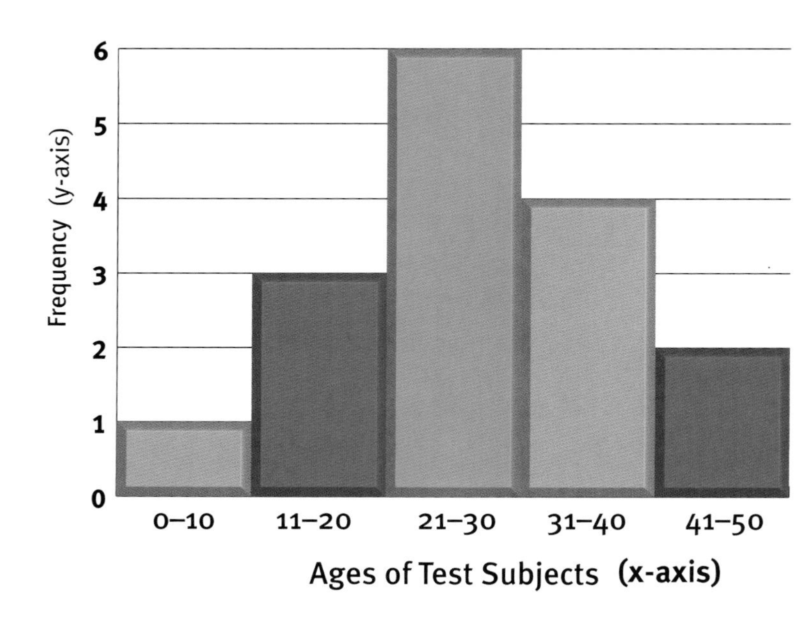

Circle Graph

A **circle graph** (SER-kul GRAF) shows how parts relate to a whole. A circle graph is also called a pie chart. This is because it looks like a pie. A circle graph has sections. Each section shows part of the whole. Circle graphs often use percentages.

Look at the circle graph below. It shows the makeup of an ecosystem. It shows the percentages for each animal group.

Circle graphs are a good way to compare the parts of a whole.

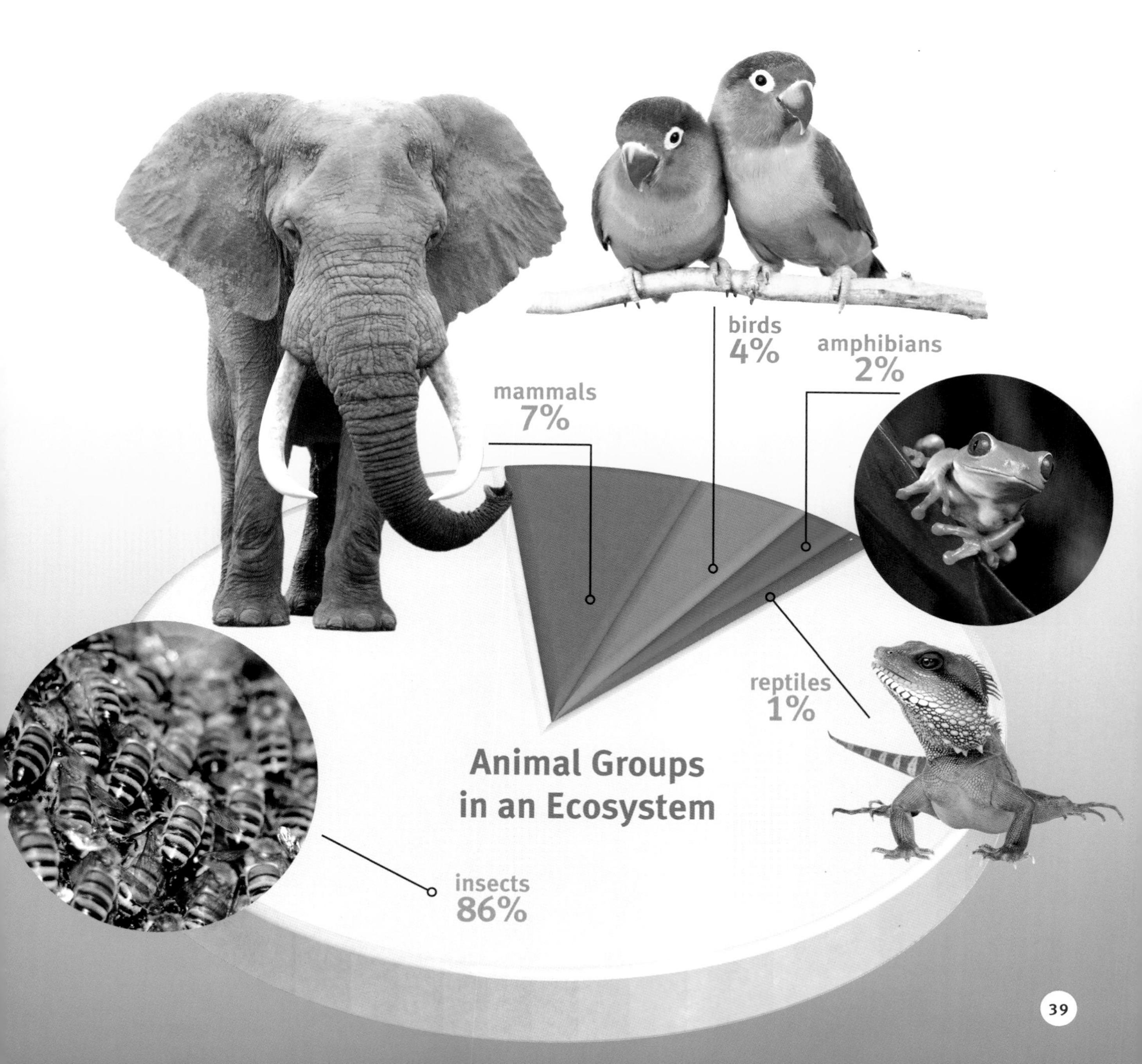

Scatter Plot

A **scatter plot** (SKA-ter PLAHT) is a way to show a pattern in a data set. A scatter plot is like a line graph. We can use this graph to plot points along an *x*-axis and a *y*-axis. But a scatter plot has many more points than a line graph. There are too many points to join with a line. These graphs are a good way to show a lot of data.

Sometimes scatter plots show a clear trend, or pattern. Other times they show no pattern at all.

When using graphs to show data, first think about the data. Then decide which graph, chart, or table would be the best way to show the data. A well-chosen graph will help people quickly understand your data.

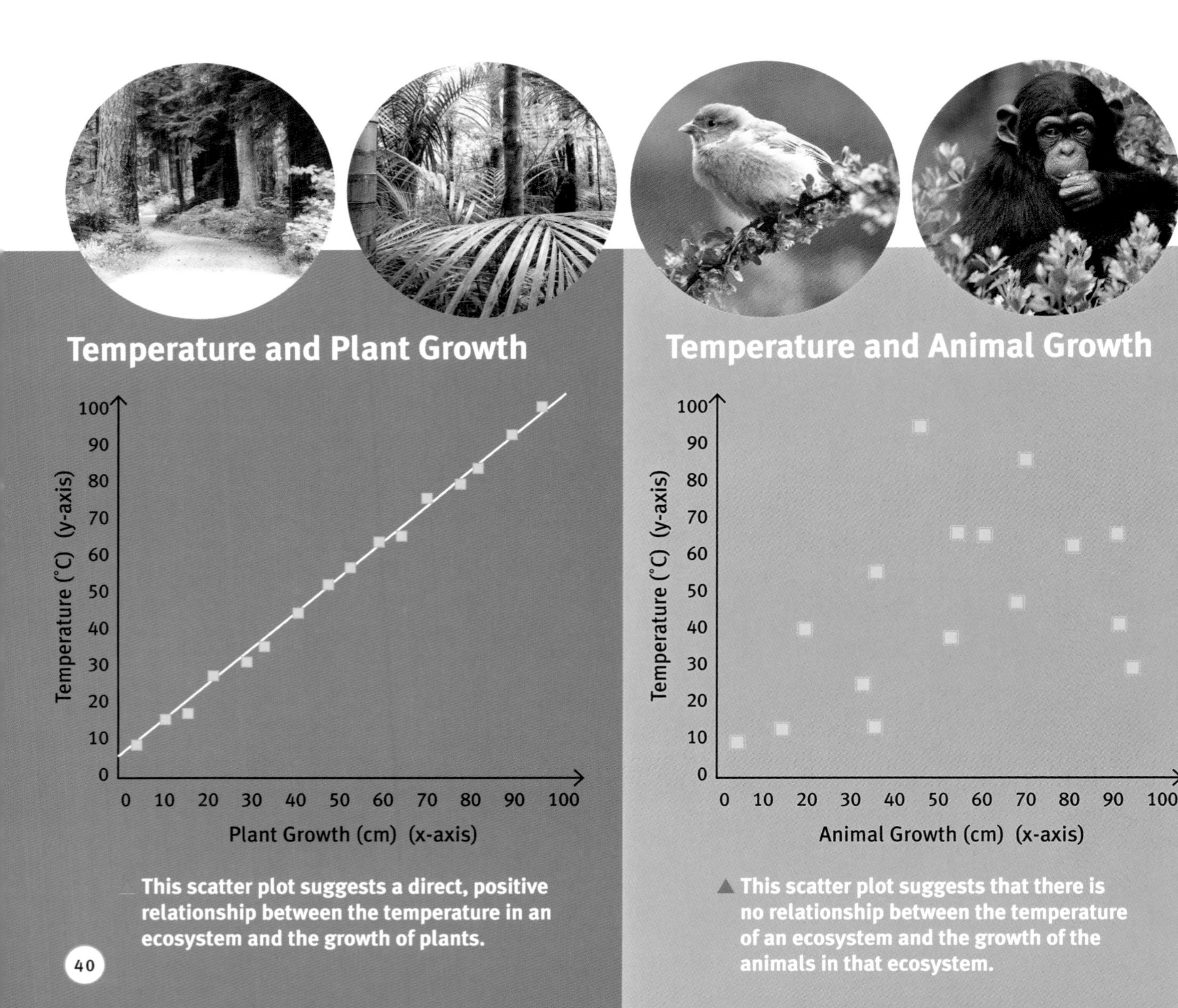

▲ This scatter plot suggests a direct, positive relationship between the temperature in an ecosystem and the growth of plants.

▲ This scatter plot suggests that there is no relationship between the temperature of an ecosystem and the growth of the animals in that ecosystem.

Summing Up

- Putting your data in a visual form is an important part of presenting it to others and analyzing it.
- Graphs can suggest trends in data and tell others where more study is needed.
- The types of graphs that scientists use include data tables, bar graphs, line graphs, histograms, circle graphs, and scatter plots.

Putting It All Together

Choose from the research activities below. Work independently, in pairs, or in small groups. Then share your findings with the class.

1. Find a bar graph in a newspaper or magazine. Attach the graph to a sheet of poster board. Present the graph to the class, identifying the information that is presented.

2. Keep track of the highest temperature where you live every day for one week in a data table. Work with a partner to create a line graph to display the data. Present your data table and line graph to the class. Describe any trends you see in the data.

3. Think of a question you can use to survey your class. For example, you might ask about favorite foods, favorite types of movies, or number of pets. In a small group, conduct your survey. Record the results in a data table. Then display the data in a circle graph.

The World of Mathematics and Science Grows

How do you picture the future world? Do you think people will live in outer space? Or do you think people will live on a greener, cleaner Earth? However you see it, we will need math and science to get us there. After all, math and science gave us the world we know today. Without math and science, we wouldn't have modern medicines. We would not have airplanes. We would not even have computer games.

Our future will also depend on math and science. Scientists around the world will do experiments. They will invent and discover new things. They will do this using math and science. They will share their findings using math and science as well. Math and science will help us make connections. Math and science will lead to new answers and new solutions. What big changes do you think math and science will bring next?

This building could not be built without the language of math and science.

HOW TO WRITE AN Objective Argument

Scientists often have to use the data from their studies to make an argument about the topic they are studying. Suppose a scientist has been studying the rain forest over a long period of time. She notices the impact that human activity has had on the ecosystem. She uses the data she has collected about the impact of human activity to support her argument that people should lessen their activity in the rain forest.

She might choose to publish her argument in a scientific journal, or in a newspaper or magazine, with the hope that the public will pay attention to this important scientific issue.

Like this scientist, you should begin your scientific argument with a clear statement of the point being made or the hypothesis being explored. The hypothesis statement should be simple and state what you are setting out to prove in your paper. The arguments you make should then be supported with measurable data and research. The more mathematical data you can present in a scientific paper or argument, the more convincing your argument will be.

Whether your argument will be read by only your science teacher or by thousands of people in the scientific community, the data and graphs you include in your paper should be clear, concise, and able to be proven through repeated experimentation or duplicated research.

1. **Choose a topic that interests you.**

2. **Gather information.**

3. **Write your argument.**
   - Present your idea.
   - State what you predict might happen if your idea is correct.
   - Present evidence/data in graphs and charts and describe how the evidence supports or contradicts your predictions.
   - Make a conclusion and propose a possible solution.

4. **Edit and revise your argument.**
   - Be careful not to give your opinion.
   - Readers will take your argument more seriously if you remain objective.
   - Be sure to carefully proofread for errors.

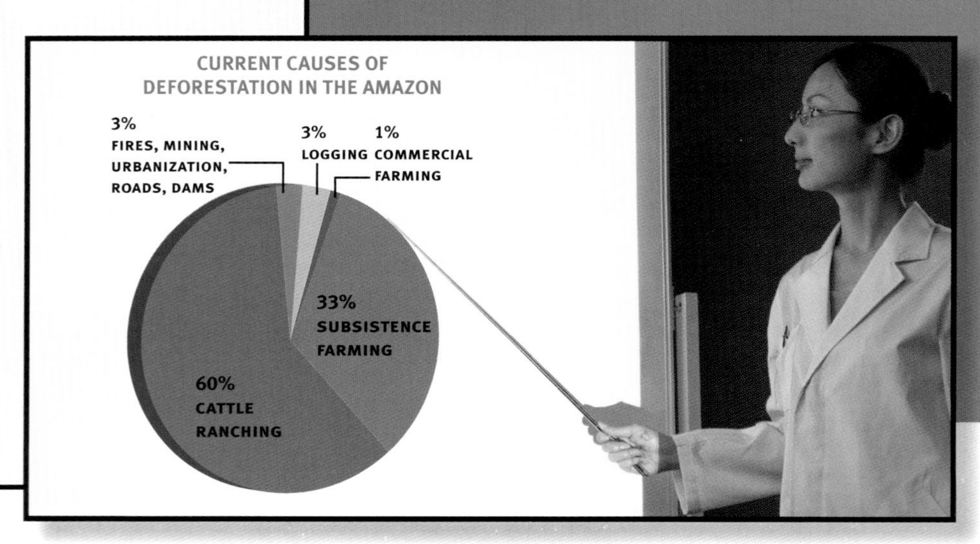

Sample Argument

The Amazon rain forest is home to more plant and animal species than anywhere else on Earth. It is also one of the world's greatest natural resources. About 20% of Earth's oxygen is produced by the Amazon rain forest. For this reason it is often called the "Lungs of our Planet." Today, 121 prescription drugs sold around the world come from plant-derived sources. Although 25% of all drugs are derived from rain forest ingredients, scientists have tested only 1% of tropical plants. Time is running out.

Today, more than 20% of the Amazon rain forest has been destroyed by human impact. The land is being cleared for ranching, mining, logging, and farming. More than half of the world's rain forests have been destroyed by fire and logging in the last 50 years. More than 200,000 acres are burned every day around the world, or over 150 acres every minute. Experts also estimate that 130 species of plants, animals, and insects are lost every day. At the current rate of destruction, it is estimated that the last remaining rain forests could be destroyed in less than 40 years.

We must work closely with other nations to reduce the level of slashing and burning. We can offer new ideas about sustainable forestry. We can also help these governments to create incentives for foreign investment such as pharmaceutical companies. Together, we can help these nations protect their own interests, and ours, by preserving this invaluable habitat, and our planet.

Glossary

bar graph (BAR GRAF) *noun* a diagram that represents number values by bars of different heights (page 35)

circle graph (SER-kul GRAF) *noun* a diagram divided into parts that represent proportions of the whole (page 39)

customary system (KUS-tuh-mair-ee SIS-tem) *noun* a system of measurement that includes ounces, pounds, cups, and gallons (page 8)

decimal (DEH-sih-mul) *noun* a fraction with a denominator of 10, or a multiple of 10 such as 100 or 1,000 (page 23)

denominator (dih-NAH-mih-nay-ter) *noun* the number below the line in a fraction; shows the number of equal parts into which the whole is divided (page 22)

dimensional analysis (dih-MEN-shuh-nul uh-NA-lih-sis) *noun* the practice of expressing units that show the same relations (page 13)

estimate (ES-tih-mit) *noun* a rough calculation that is not exact (page 16)

exponent (EK-spoh-nent) *noun* a smaller number that is placed after and above another number to show how many times that number is to be multiplied by itself (page 14)

formula (FOR-myuh-luh) *noun* a mathematical rule or relationship that is expressed in symbols (page 12)

fraction (FRAK-shun) *noun* a number that is a part of a whole (page 22)

histogram (HIS-tuh-gram) *noun* a graph that shows how data is distributed in a set (page 38)

improper fraction (im-PRAH-per FRAK-shun) *noun* a fraction with a numerator larger than the denominator (page 22)

line graph (LINE GRAF) *noun* a diagram that represents the relationship between two variables (page 36)

mean (MEEN) *noun* the sum of the numbers in a data set divided by the number of items in the data set (page 30)

median (MEE-dee-un) *noun* the middle number in a data set (page 30)

metric system (MEH-trik SIS-tem) *noun* a standardized system of measurements based on the number 10 (page 9)

mode (MODE) *noun* the most frequently occurring number in a data set (page 30)

numerator (NOO-muh-ray-ter) *noun* the number above the line in a fraction; shows how many parts of the denominator are taken (page 22)

percent (per-SENT) *noun* the number of parts in every 100 (page 23)

proportion (pruh-POR-shun) *noun* a relationship between two ratios (page 25)

range (RANJE) *noun* the difference between the lowest and highest numbers in a set (page 30)

rate (RATE) *noun* a type of ratio that presents two terms in different units (page 24)

ratio (RAY-shee-oh) *noun* a relationship between two quantities that is expressed as the quotient of one divided by the other (page 24)

scatter plot (SKA-ter PLAHT) *noun* a diagram that shows two variables that may or may not suggest a relationship between them (page 40)

scientific notation (sy-en-TIH-fik noh-TAY-shun) *noun* a way of writing numbers in a decimal form by using exponents (page 14)

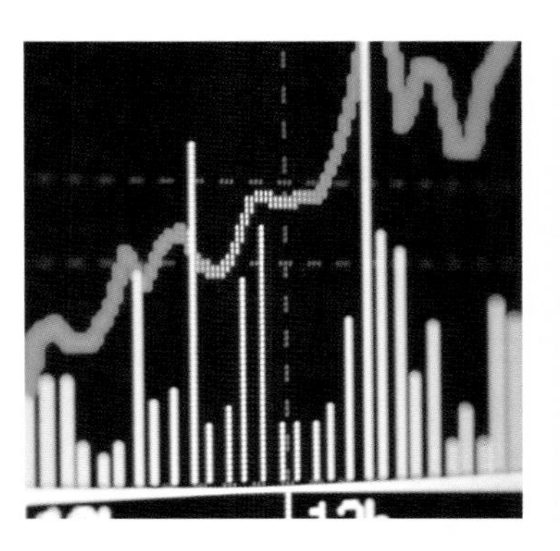

Index